iScience
Readers

Solids and Liquids:
Who Messed Up My Sand?

by Emily Sohn and Joel

Chief Content Consulta
Edward Rock
Associate Executive Director, National Science Teachers Association

NORWOOD HOUSE PRESS
Chicago, IL

Norwood House Press
PO Box 316598
Chicago, IL 60631

For information regarding Norwood House Press, please visit our website at
www.norwoodhousepress.com or call 866-565-2900.

Special thanks to: Amanda Jones, Amy Karasick, Alanna Mertens, Terrence Young, Jr.

Editors: Barbara Foster, Diane Hinckley
Designer: Daniel M. Greene
Production Management: Victory Productions

Paperback ISBN: 978-1-60357-282-8

The Library of Congress has cataloged the original hardcover edition with the following
call number: 2010044528

Manufactured in the United States of America in North Mankato, Minnesota.
278R—042015

Contents

Note to Caregivers:

Throughout this book, many questions are posed to the reader. Some are open-ended and ask what the reader thinks. Discuss these questions with your child and guide him or her in thinking through the possible answers and outcomes. There are also questions posed which have a specific answer. Encourage your child to read through the text to determine the correct answer. Most importantly, encourage answers grounded in reality while also allowing imaginations to soar. Information to help support you as you share the book with your child is provided in the back in the **Additional Notes** section.

Words that are **bolded** are defined in the glossary in the back of the book.

Playing in sand is fun!
But what if the sand
isn't just sand?

4

How Can You Get the Salt Out?

Your sandbox needs more sand.

Oops, someone added a big bag of salt instead of sand!

The **grains** of salt and sand are about the same size.

How can you get the salt out?

Salt in the Sandbox

Here are three ideas for getting the salt out:

Idea 1: Pick the salt out of the sand with your fingers.

Idea 2: Pour the salt-and-sand **mixture** through tiny holes in a screen **filter.** Maybe the grains of salt or sand will get stuck and stay on top, and the other grains will go through the holes.

Idea 3: Stir the salt and sand into a bucket of water. Then, pour the water mixture through the tiny holes in a screen.

kitchen sieve or sifter

Which idea do you think will work best?

Need help? Try the Discover Activity.

Discover Activity

Materials

- sesame seeds
- dried kidney beans
- kitchen strainer

Sorting by Size

One way to put things in groups is by size.

Grouping is also called sorting. Try this.

Mix beans and sesame seeds in a strainer.

A strainer is a bowl with holes in it.

sesame
seeds

beans

strainer

What will happen?

The beans are big. They will stay in the bowl.

The sesame seeds are small. They will fall through the holes.

You have sorted the foods by size.

salt

sand

The salt and sand are shown larger than actual size.

Could you use a strainer to get the salt out of your sandbox?

Look closely at the pictures of salt and sand.

The grains are about the same size.

Sort This

How would you sort these toys?

When you sort, you look at how things are the same. You also look at how they are not the same. These things you can sort by are called **properties.**

Size is a property. So is **hardness.**

A toy can be big or small. It can be soft or hard. Can you think of other ways to sort things?

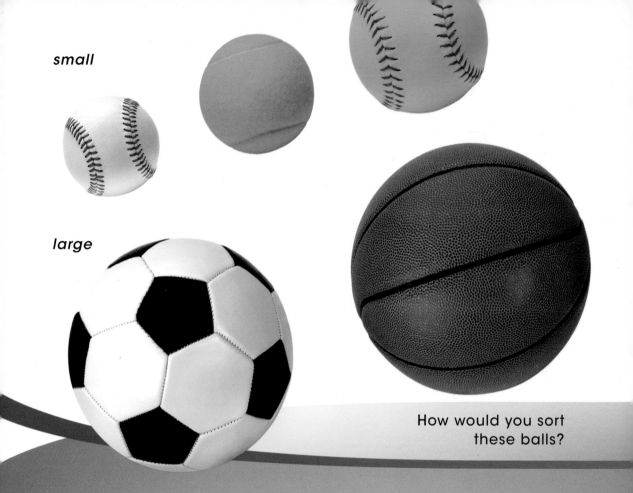

small

large

How would you sort
these balls?

In the picture, these balls are sorted by size.

How else could you sort them?

Which balls are soft? Which balls are hard?

What color are they? Do some have more than
one color?

A mortar and pestle are tools used to crush, grind, and mix solids.

Sand and salt have some things in common.

They are both hard. Both are also **solid.**

What makes them not the same?

Pick up a rock and try crushing a chunk of sand with it.

Then try crushing a chunk of salt with it.

What happens to the sand and the salt?

Solid as a Rock

Like salt and sand, you are a solid. A solid does not change shape by itself.

A solid can be moved. But it does not flow like water.

Solids can be hard or soft.

Some solids are easy to break. Others are really strong.

Which of the solids in the picture break easily?

Go with the Flow

Water is a **liquid.** Liquids have different properties than solids have.

Liquid splashes. When you pour it, it changes to the shape of what you put it in. You always look like you. But water takes the shape of what you put it in.

This girl is pouring water on the sand.

What do you think will happen next?

Solids

- have their own shape.
- can be any color.
- can be different **textures.**

Liquids

- take the shape of their container.
- can be any color.
- can be thick or thin.

What happens when
you stir salt into water?

Water is clear. But it can still have stuff in it. What happens when you stir salt into water?

The salt breaks into tinier pieces. It **dissolves.** You can't see it anymore. But it's there.

Salt is **soluble.** Not all solids dissolve in water.

What happens if you put a crayon in water?

Sorting Tricks

Another way to sort solids is to see whether they dissolve.

Which of the solids below will dissolve in water?

Which will not dissolve?

sugar

gravel

beads

Let's think about the sandbox problem again.

You can try pouring water into your sandbox.

What do you think will happen to the salt?

What will happen to the sand?

Will just one dissolve? Will both dissolve?

People used the properties of solids and liquids to pan for gold.

Connecting to History

Panning for Gold

In 1849, people found gold in California. The gold was mixed with dirt at the bottom of rivers.

How did people sort the gold from the dirt? First, they swished water and dirt around in a pan. The pan had little holes in the bottom. Lots of stuff fell through the holes. Only gold dust and heavy dirt stayed in the pan.

Then it was easy to pick out the shiny gold.

Knowing about solids and liquids helped these people solve a problem!

Vanishing Act

Ocean water tastes salty.

That's because salt dissolves in water.

It breaks into smaller pieces.

Stir salt into a glass of water. What happens?

So mixed together, salt and water can go through tiny holes in a screen.

Sand does not break into pieces in water. It stays as a solid. It does not dissolve.

whistling sand dune in Dubai

Did You Know?

Some sand whistles or sings! Here's how: Sand builds up in dunes and on beaches. Each grain of sand has a special shape. When wind blows or people walk on it, the sand makes a noise. Now, that's music to our ears!

Artist

An artist paints a picture. On the tip of a paintbrush, the paint is a liquid. When the paint dries on the paper, it is a solid. Why? Paint is really a mixture. It is water with solids dissolved into it. When paint dries, just the solids are left behind.

Paint can be a liquid or a solid or both.

How good are the ideas?

Idea 1: Not great. It would be a lot of work to pick out each grain of salt.

Idea 2: Not good. Both salt and sand would stay in the screen.

Idea 3: Pretty good! The salt would dissolve. Then the salty water would go through the holes. The sand would not.

Now, you can scrape the sand off the screen. You can put it back into your sandbox!

Beyond the Puzzle

In this book, you learned how to sort solids.

You also learned how to sort a mixture.

Now imagine going to a beach to collect shells.

It is hard work to pick shells out of the sand, one by one.

Can you think of a better way to collect lots of shells without bringing home the whole beach?

How would you sort the seashells from the sand?

Glossary

dissolves: seems to disappear in a liquid.

filter: a solid with holes that block big pieces but let small pieces and liquids pass through.

grains: tiny pieces of sand or salt.

hardness: how difficult it is to break through the outside of a material.

liquid: something that can be poured.

mixture: two or more things stirred or shaken together.

properties: words that help describe an object.

solid: something that has a definite shape.

soluble: able to be dissolved in a liquid such as water.

textures: how surfaces of materials feel.

Further Reading

Looking at Solids, Liquids, and Gases: How Does Matter Change?, by Jackie Gaff. Enslow Elementary, 2008.

Sort it Out!, by Barbara Mariconda. Sylvan Dell Publisher, 2008.

The Wild Water Cycle, by Rena Korb. Magic Wagon, 2008.

Additional Notes

The page references below provide answers to questions asked throughout the book. Questions whose answers will vary are not addressed.

Page 8: No. The salt pieces are about the same size as the sand pieces. The strainer could not separate the sand from the salt.

Page 11: The salt will break apart into smaller pieces. The size of the pieces of sand will not change.

Page 13: The water will sink into the sand.

Page 14: The crayon stays the same shape. It does not dissolve.

Page 15: Sugar will dissolve, but gravel and beads will not. The salt will dissolve, but the sand will not.

Page 17: Caption question: The salt dissolves.

Index